des&desi

Chemistry In My DREAMS

BOOK 2: MAN-MADE PROCESSES

WRITTEN BY TERRESSA BOYKIN
Illustrated by Billi French

Copyright 2017 by The Boykin STEM Inc.
All rights reserved. This book or any portion thereof may not be reproduced or used in any manner whatsoever without the express written permission of the publisher except for the use of brief quotation in a book review.

ISBN: 978-1-945751-03-5

Library of Congress Cataloging-in-Publication Data
2017906375

Boykin, Terressa
des&desi
Chemistry In My Dreams

TABLE OF CONTENTS

Prologue
In My Dreams 1

Punchy Juice Boxes 7

The Icy Popsicle Pool 17

Epilogue
In My Dreams 25

This book is dedicated to
Larry E. Collins Jr. for igniting the spark
for my love of chemistry and to
James V. Mason for cultivating the flame.

PROLOGUE
In my dreams

Each night when I go to bed
To a dreamland in my slumber,

With drooping eyes I slip into
The learning land of wonder.

In my dreams I see a wonderland
Of how and what things do

And how we learned our chemistry
In elementary school.

PUNCHY JUICE BOXES

An Introduction to the Compatibility and Incompatibility of Liquids.

**Two juice boxes met at lunchtime
And had very bad behavior.
They didn't like each other,
For each had a different flavor.**

8

They would poke and poke each other,
Whenever they would meet.
And all because one was lemon
And the other was cherry sweet.

One day they drew their straws
And hit with all their might.
When suddenly – *POP* – a leak began
To end their fruitful fight.

10

As a puddle grew between them
On the table during lunch,
"Wow!" said one, "When mixed together,
We make a tasty punch."

**The juice boxes are now friends,
And together they have found
That things can be much better
When the other is around.**

We see mixtures can be made
From the liquids that we drink,
Even those that may not seem
As well-matched as we might think.

Other liquids may be mixed
Without putting up a fight,
But may become more difficult
When they are not alike.

**With time and a little energy
The two can become one
While carefully observing them
To see what they become.**

THE ICY POPSICLE POOL

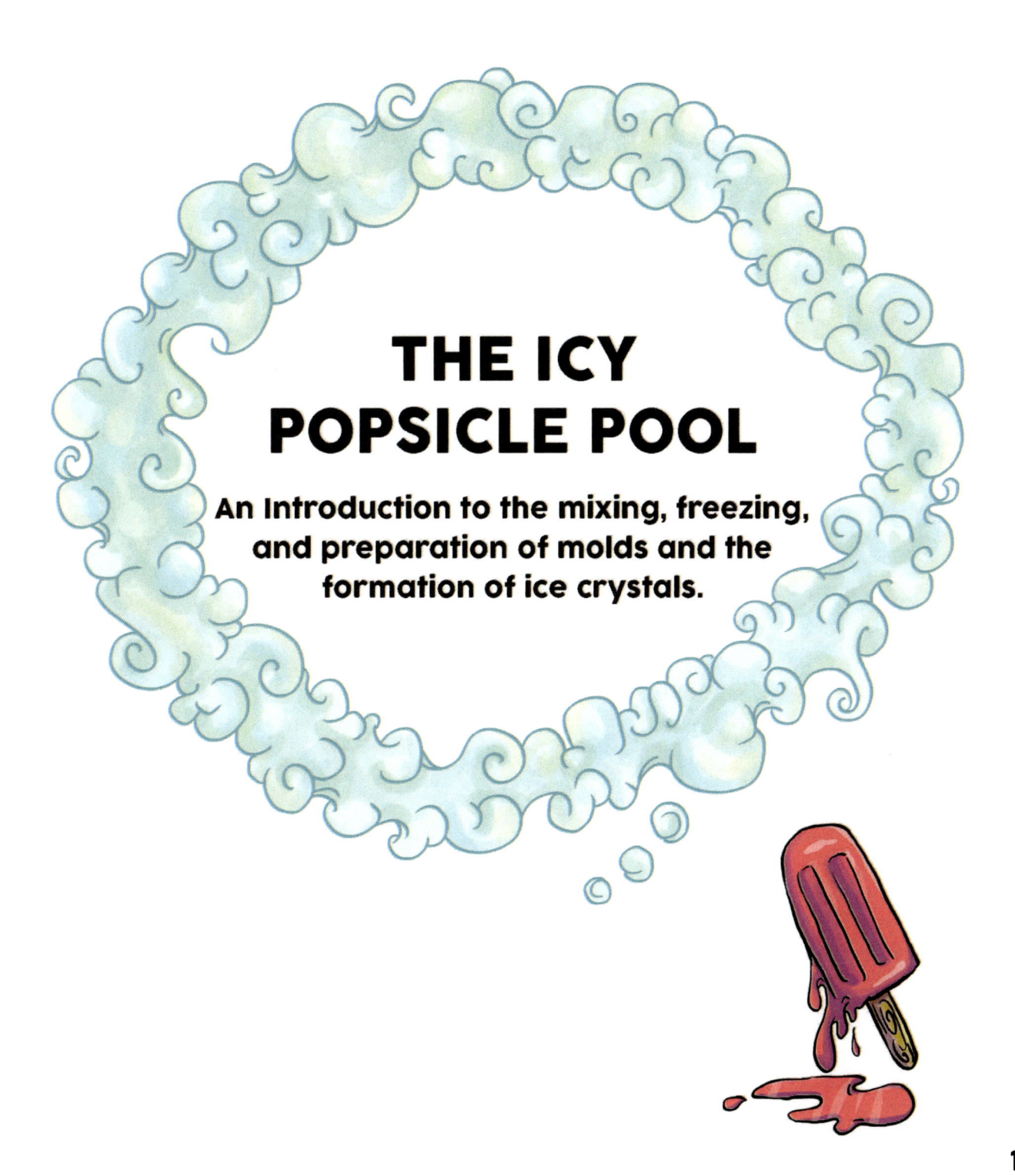

An introduction to the mixing, freezing, and preparation of molds and the formation of ice crystals.

**With crushed fruit and nectar juice,
A popsicle you can make.
Bringing these things together
Is the first step you should take.**

**Now, if you now stir the mixture
You will see how it unfolds,
While pouring into containers
Of any shape to make the molds.**

**If you freeze the liquid mixture
For a few hours in the cold,
Water crystals slowly form
Wrapped in colors, bright and bold.**

**The frosted chips begin to glitter
With fruit juices shining through,
To make brightly colored popsicles
Of ice crystals and fruit juice.**

**The chill and ice will wet your tongue,
Too cold to take a nibble,
But if you wait a minute or two
It will melt into a dribble.**

See if you can guess
If it is grape, raspberry or cherry.
So close your eyes and slurp it up
And taste the fruity berry.

EPILOGUE
In My dreams

Snuggling with my pillow
In this enchanted place,

**Quietly, I am sleeping
With a smile upon my face.**

In my dreams I see a wonderland
Of how and what things do.
All we learned of chemistry
In elementary school.

The night is almost over
To another day I'll wake.
But, until then I'll keep dreaming,
More to see, new friends to make.

THE END.

ACKNOWLEDGEMENTS

ABOUT THE AUTHOR:
Terressa Boykin holds a Masters of Science in Organic and Kinetic Chemistry and has worked in the field of Polymer Chemistry and Chemical Engineering for over 28 years. In addition to volunteering in her community as a local and regional science fair judge, she is a member of the American Chemical Society's Coaching Program, the Education and STEM Committee for her local Chamber of Commerce. She is the founder and CEO of Des&Desi Chemistry Classes for young children and the author of the children's book series Des&Desi *Chemistry In My Dreams, The Atom: What Am I Really?,* and *Matter: Is All Around You.*
www.desndesi.com

ABOUT THE ILLUSTRATOR:
Billi French has a Bachelor of Arts Degree in Studio Art and a Certification in Illustration and Animation. She is the owner of the illustration company, BF Studios.
www.billifrench.com.

CONCEPTUAL ARTIST:
Donna Merchant holds a Bachelor of Fine Arts Degree, majoring in art education. She is a resident artist/art teacher at Red Art & Design Studio.
www.redartanddesign.com

MY INSPIRATION:
Desmen A. Boykin is a middle school student who has a passion for science and a talent for creative drawing. He hopes to be an engineer one day.

Desiree M. Mosley is an LPN and currently an RN nursing student who has a gift and a passion for helping others.

Other books by the author

DES&DESI
THE ATOM: WHAT AM I REALLY?

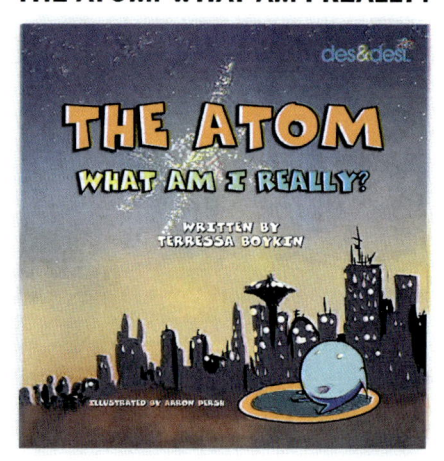

DES&DESI
MATTER: IS ALL AROUND YOU

DES&DESI
CHEMISTRY IN MY DREAMS
BOOK 1: NATURAL PROCESSES

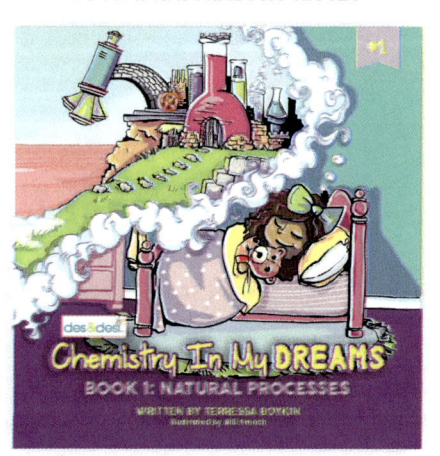

DES&DESI
CHEMISTRY IN MY DREAMS
COLORING BOOK 1: NATURAL PROCESSES

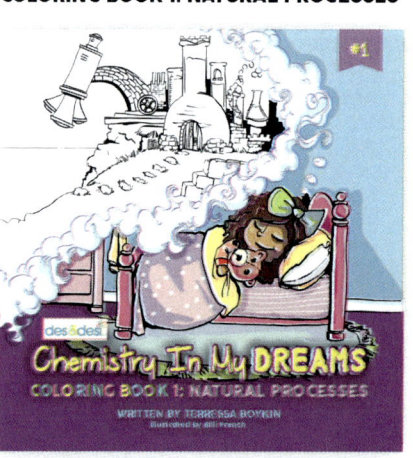

DES&DESI
CHEMISTRY IN MY DREAMS
COLORING BOOK 2: MAN-MADE PROCESSES

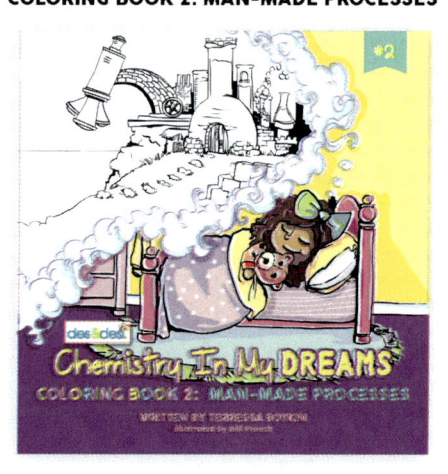

Made in the USA
Middletown, DE
14 December 2021